BIOHERBICIDE

BENUKAR BISWAS

M. Sc, Ph. D.

Associate Professor of Agronomy

Directorate of Research, Bidhan Chandra Krishi Viswavidyalaya, Kalyani, Nadia, West Bengal-741235

E-mail: kripahi@yahoo.com

Dedicated to

Late Dr S. K. Mukhopadhyay, Ex Vice

Chancellor, Viswa-Bharati University

Dr L S Brar, Ex Professor, Punjab

Agriculture University and

Late Dr R D Singh, Scientist, Indian

Institute of Himalayan Biodiversity, CSIR

CONTENTS

PARTICULARS	PAGE NO
1 INTRODUCTION	5

APPROACH OF BIOLOGICAL CONTROL
BIOHERBICIDE

| **2 MYCOHERBICIDE DEVELOPMENT** | 10 |

COLLABORATIVE EFFORT

PROCESS OF DEVELOPMENT

CRITERIA FOR SELECTION OF POTENTIAL

MYCOHERBICIDE

COMMON RESTRAINTS ON PATHOGEN

EPIDEMIOLOGICAL MECHANISMS OF

     MYCOHERBICIDE EFFECTIVENESS

| **3 CONSTRAINTS IN THE** | 15 |

DEVELOPMENT OF BIOHERBICIDE

BIOLOGICAL CONSTRAINTS

ENVIRONMENTAL CONSTRAINTS

TECHNICAL CONSTRAINTS

COMMERCIAL LIMITATIONS

| **4 FORMULATIONS** | 27 |

PARTICULARS	PAGE NO

5 PROMINENT MYCOHERBICIDES   31

DEVINE

COLLEGO

OTHER PROMINENT MYCOHERBICIDES

6 PROSPECT   34

REDUCTION OF GROWTH AND
     COMPETITIVENESS - AN ECOLOGICAL
     APPROACH

INCLUSION OF MYCOHERBICIDES IN IPM

PROGRAMME

7 ECONOMICS   39

8 SUMMERY   40

REFERENCES   41

Table 1 Prominent mycoherbicides   73

Table 2. A sampling of pathogens   74

used, in the market or amid study

1   INTRODUCTION

In integrated weed management system, herbicides are used at minimal levels to keep pesticides' residues as least as possible to minimize their adverse effect on environment. Wide spread occurrence of herbicide resistance and cross resistance (isoproturon resistant *Phalaris minor* in wheat, Holt *et al.* 1990 and Moss *et al.* 1993), herbicide persistence (picloram and atrazine in air, Grover 1991; in soil - Smith 1986; and in water Lym *et al.* 1988; bromyxnil, carbmate and diuron in fish through water) and problems of spray drift are serious issues of chemical weed control. In UK, Bals (1984) estimated 72 million litre of spray liquid as long distance drift per year. Only 5% of applied herbicide is retained by weeds (Combellack 1981 and Pimental *et al.* 1986). Besides these factors, the growing concern regarding pesticide's residues and environmental degradation has made imperative to evaluate alternate system of weed management involving biological approaches to combat weeds.

APPROACH OF BIOLOGICAL CONTROL

There have been mainly two approaches of biological control *viz*. Classical and Inundative.

In classical or inoculative approach, an exotic biocontrol agent is introduced in a small area of infested zone e.g. induction of a beetle *Dactylopus* sp. into Australia to control *Opuntia* sp. (Gressel *et al.,* 1996). In this approach, the control is slow and also dependent on favourable ecological conditions promoting an epidemic and causing gradual increase in attack or infection of plants. Manipulation of increasing efficacy may be difficult or impossible and there is increasing concern over the potential risk of introducing exotic organism. It has been employed in controlling weeds of range-lands, water-ways and semi-natural areas (extensive agriculture) but it has limited success in intensive agriculture.

On the other hand, inundative or space bioherbicide approach has been employed to control indigenous weed species with native pathogen applying the bio-agent (mainly fungus) in massive dose in the area infested with target weed flora (Muller Seharer *et al.,* 1996).

BIOHERBICIDE

There seems to be some controversy regarding definition of a bio-herbicide. According to Watson (1989), bio-herbicides are living entities (natural enemies) used deliberately to suppress the growth or reduce the population of a weed species. This may include an insect, a microbe or a parasite nematode. These organisms operate directly or indirectly (by producing toxins) and their mode of action proves deleterious to the weed species. The latter toxins are termed bio-rationals, phytotoxins or natural herbicide and are excluded in the definition.

Bio-herbicides are formulations containing plant pathogen capable of mass production *in vitro* that are applied directly to target weeds uniformly to kill or suppress the growth of weed. This method employs the massive, usually annual release of a bio-agent into specific weed infested fields to inflect and kill susceptible weeds. At present, only the potential use of fungal pathogens as myco-herbicide is being studied in depth. Exploitation of bacteria seems promising, whereas use of virus as biocontrol agent is problematic since they are not host specific and require vector for their transmission.

Commercial bioherbicides first appeared in the market in USA in early 1980s with release of the product Devine in 1981 (Kenney 1986) and Collego in 1982 (Bowers

1986) and Biomal in Canada (Makowski 1998). Success story of these products and expectation of getting perfect analogues of chemical herbicides have opened a new vista of weed management. Purpose of this review is to overview its development, constraints associated, formulations discovered, present status and future scope, economics and overall research directions in this regard.

2 MYCOHERBICIDE DEVELOPMENT COLLABORATIVE EFFORT

The mycoherbicide development process involves multidisciplinary approach and dedicated co-operation among weed scientist, pathologist, food technologists or fermentation specialist and industry (Raj Prasad 1993).

PROCESS OF DEVELOPMENT

The process of development of an effective bio-herbicide has been found to have three steps (Raj Prasad 1993).

The discovery phase: The usual scenario begins with the discovery of diseased weed species in the field; the causal agent is then cultured on artificial media, identified and re-inoculated back into the host weed. If the disease is reproduced, the same agent is re-isolated, the fungus is further tested as a potential mycoherbicide.

The development phase: It includes testing of infectivity and pathogenicity under different environmental conditions, small scale field tests, host-range studies, studies on fungus survival in the field, test of compatibility with chemicals and other treatments and evolution of herbicidal efficacy. Based on these indications, efforts are made for development of mass culture techniques, devising suitable formulation and application technology, testing for allergenic or toxic effects on animals and further tests of efficacy, non-target effects and environmental fate (Charudattan *et al.* 1982).

The commercialization phase: This ultimate phase begins after successful registration usually in co-operation with a private industry leading to mass production and marketing.

CRITERIA FOR SELECTION OF POTENTIAL MYCOHERBICIDE

Daniel (1973) listed general requirement for selection of potential bio-herbicide. He considered that pathogen must be able to produce abundant and durable inoculum in artificial culture, be genetically stable and specific for the target weed, and be able to infect and kill the weed in environment of reasonably wide latitude. While selecting potential bio-herbicide, high parasitism between saprophytes and obligate parasites was also given importance (Templeton 1982).

COMMON RESTRAINTS ON PATHOGEN

Although till date a good many number of fungus have been reported as potential mycoherbicide, but only a few were released as commercial product. Many of these fungus failed a few tests and were discarded. Common problems associated with pathogen were reported to be poor over-wintering capacity, poor inoculumn dissemination, requirement of long incubation period, low virulence and host resistance, inadequate host density and fastidious environmental parameters or sequences required by the pathogen for its efficacy.

EPIDEMIOLOGICAL MECHANISMS OF MYCOHERBICIDE EFFECTIVENESS

Epidemiological studies regarding mycoherbicide have indicated importance of temperature, moisture or dew period requirement, inoculumn concentration and plant growth stage influencing initial infection.

Makowski (1993) reported highest level of control at inoculum concentration of 2×10^6 spores/ml of *Colletotrichum glocosporioides* sp. *malvae* on younger seedlings of weeds *Abutilon theophrasti* and *Malva pusilla*. Optimum temperature and minimum dew requirement has been reported to be 20-25°C and 20-48 hours, respectively. The epidemiological mechanism was considered to be over with the subsequent dispersal and infection of the pathogen on the target weed population (Yang *et al.*, 1993).

3   CONSTRAINTS   IN   THE DEVELOPMENT OF BIOHERBICIDE

Bioherbicides have certain advantages. They are highly host-specific, eco-friendly and beneficial to nontarget species, free from chance of resistance development, non-persistent, cost effective, very promising in controlling resistant weeds, mimic weeds, parasitic weeds, but certain factors have been reported to limit the development of bio-herbicides into a commercial product. Bruce and Louise (1995) have classified those factors under four categories.

BIOLOGICAL CONSTRAINTS

Plant populations. - Host variability and host range: Within a population of host weed species there will usually be a range of genetically diverse biotypes (Burdan 1987) which may include same resistant types. It is also usual of having different biotypes of micro-organism (Weidemann *et al.* 1990) with slightly different host ranges (Auld *et al.,*

1992, Boyette 1991, Mikandrow *et al.*, 1990, Schepens 1987). So, there is chance to mixing within a fungal species used as a bioherbicide. There is some evidence of a lack of clear relationships between plant phylogamy and pathogen species (Walter *et al.* 1991, and Weidemann 1991) making choice of species used in complex host range tests difficult. Besides, concerns have been raised in recent years regarding the potential for sexual and asexual gene exchange between bio-herbicide strains and strains attacking distantly related crop plants (TeBeest *et al.* 1992 a, TeBeest *et al.* 1992b, Weidemann 1992) and natural exchange of genetic material between pathogen (TeBeest *et al.* 1992 a). Successful mating in culture and plant tissue of *Colletotrichum glocosporioides* f. sp. *aeschynomene,* the active ingredient of Collego, with strains of some fungus infecting Winged water primrose *(Ludwigia decemens)* and *Carya illinoensis* (TeBeest *et al.* 1992 a).

Individual Plants: Resistance mechanisms and assessment. There are myriad barriers to infection on individual plants, some inherent, some induced; including physical barriers hair and cuticle; biological barriers such as antagonistic phylloplane, micro-organisms; and chemical barriers such as phytoallixins (Deverall 1977, Misaghi 1982).

Individual plants have been reported to differ in their ability to tolerate low levels of infection and exhibit no macroscopic reaction (Auld *et al.* 1988, Morin *et at.* 1989 and

Vanderplank 1975). The age (Makowski 1993 and Auld *et al.* 1990) phenological stage, physiological stages (Schoenewiess 1975) of plant also have been found to influence the host-pathogen-environmental triad. For instance, spiny cocklebur *(Xanthium spinosum* L) found relatively resistant to *Colletotrichum biculare* (Berk & Mont) V. Arx at moisture stress conditions (Auld *et al.* 1990). Morphological character e.g. position of buds, growth habit (Charudattan *et al.* 1990, Rounkiar 1983) were also considered to determine the efficacy of a bioherbicides.

Some plants were observed to recover quickly from disease symptoms (Makowski 1993, McRae *et al.* 1988, Morin *et al.,* 1993 and Wymore *et al.,* 1988) and thus assessment of the potential of microorganisms as bioherbicide became difficult. Charudattan reviewed assessment methods, pointing out the difficulty of discriminating between candidates that had sublethal effects.

Interaction with other microorganisms: The presence of microbial competition for nutrients and space or direct antagonism occurring in the plynosphere (Blakeman 1992, Kenerley *et al.,* 1990 & Windels *et al.,* 1985) as well as toxic plant leachates (Kenerley *et al.,* 1990 and Tukey 1970) were observed to reduce the efficacy of foliar bioherbicide.

ENVIRONMENTAL CONSTRAINTS

Epidemiology was also dependent on optimum environmental condition (TeBeest

1991). In the field, a high number of initial infection sites were generally essential for weed control (Yang *et al.* 1993)

Aerial environment : The prevailing environmental conditions were found to be limiting factors for biological control agents especially for fungi (Andrews 1992 & Kenerley *et al.* 1990), for example requirement of particular dew period (Boyette *et al.*, 1984), Macaque 1993, McRae *et al.*, 1988, Morin *et al.*, 1990, Wymore *et al.*, 1988), humidity (Auld *et al.*, 1990, Makowski 1993, McRae *et al.*, 1988, Soul *et al.*, 1992 and Templeton 1992) and temperature (Auld *et al.*, 1990, McRae *et al.*, 1988 & TeBeest *et al.*, 1992b).

Soil environment: Moisture and nutrient status of the soil were of target plants and their interaction with aerialy applied bioherbicide. Soil reported to influence bioherbicide efficiency which affects the physiology borne pathogens applied at pre-emergence stage of weeds were reported to tolerate environmental extremes and became persistent and effective to residual control (Boyette *et al.*, 1984, Jones *et al.*, 1990 & Widemann 1988) but they usually had high virulence on a wide range of plant species (Jones *et al.* 1990 & Feicthenberger *et al.*, 1984).

Aquatic environment: Charudattan *et al.*, (1990) have discussed the technical and environmental limitations regarding application of bioherbicides and maintaining their efficacy in water. They have listed the limiting factors to be oxygen concentration, temperature, salinity and diffusion of spore in water.

TECHNICAL CONSTRAINTS

Mass production : Submerged fermentation technology used in case of Devine, Collego and Biomal (Boyette 1991, Cunningham *et al.,* 1989 and Kenney 1986) were not cost effective. Large scale solid substrate fermentation technology which has been available in industrialized countries (Christy *et al.,* 1993) is unfortunately not available for mycoherbicide for mass production (Connick *et al.,* 1990 and Stowell 1991). Selection or design of the growth medium (Bannan *et al.,*

1990, Jackson *et al.,* 1990, 1992, Morin *et al.,* 1990b and Schister *et al.,* 1991), improvement of efficiency of submerged fermentation and stabilization process for propagules, and selection of type of fermentation systems (Connick *et al.,* 1990) were also suggested to be considered.

Formulation: Inadequate formulation has been one of the major constraints to the development of reliable and efficacious bioherbicide (Greaves *et al.,* 1992). Liquid formulation of Devine has a very short shelf-life (6 weeks) (Kenney 1986) limiting market area. Most challenging aspect of formulation has been to overcome the dew requirement which has also reported to reduce the dosage of inoculumn and cost. In this direction, invert

24

emulsion were found to have certain potential (Connick *et al.* 1991, Daigle *et al.* 1990 and 1992 and Quimby *et al.* 1989). However, some non-target damage has been reported due to its high viscosity (Auld 1993 and Womack *et al.*, 1993). Connick *et al.*, (1991) have reported absence of dew requirement developing an invert emulsion formulation of lower viscosity and higher water retention property.

COMMERCIAL LIMITATIONS

Market size: Several efficient bioherbicide candidates remained underdeveloped for commercial use by industry because of their low market potential (Charudattan 1991, Templeton 1982, 1992). Bioherbicide for economically important weeds (Charudattan *et al.*, 1992, Greaves *et al.*, 1992 , Heiny 1993 and Zorner *et al.*, 1993), opportunities of cost recovery (Templeton 1992), size of foreseeable markets, availability of alternative chemical herbicides (Charudattan 1991) have been the determining factors of market size.

Patent protection and secrecy: The aim of patent system is to promote science by recognizing and rewarding inventors but it has been found to delay the exchange of ideas and results among scientists (Saliwanchick 1986, 1988).

Regulation: Increasingly restrictive and often unrealistic regulations for bioherbicide in North America have raised the cost of registration and discouraged investors

(Makowski 1993). There is growing concerns amongst regulators regarding the possible production of mycotoxins by fungus e.g. *Phomopsis emieis, Fusarium* spp. used as bioherbicide (Beier 1990, Hesseltine 1986 and Makowski 1993)

4   FORMULATIONS

Goal of formulations is to keep propagules viable and effective for a reasonable amount of time (shelf life).

TYPE - Following types of formulation have been developed commercially and at the laboratory scale.

Liquid formulation: Liquid formulation of Devine has been developed. But it requires particular dew period for its efficacy.

Invert emulsion: It consists of oil suspended in water which has been observed to retard evaporation of water spray droplets and replace or minimize the dew requirement and help in producing microclimate for disease development. Connick *et al.,* (1991) has reported an invert emulsion of low viscosity and high water retention of *Alternaria cassiae*, a potential pathogen against sickle pod (*Cassia obtusifolia*) Boyette *et at.,* (1993) has also reported effective biological control of Hemp Sesbania under field conditions with *Colletotrichum truncatum* formulated in invert emulsion.

Granular *formulation:* Encapsulation of propagules within cross linked matrix organic polymers (e.g. sodium alginate) has been investigated in great details (Daigle *et al.,* 1992, Walker 1981, 1983). Connick *et al.,* 1991 have reported the encapsulation of fungal propagules in a wheat gluten matrix following the process employed for making pasta products. These pasta formulation was reported to be effective using many fungus to control respective target weed e.g. *Fusarium oxysporum*, to control sickle pod, Coffea sinna (*Cassia occidentalis*) and Hemp Sesbania (Boyette *et al.,* 1993); *Alternaria cassiae* to control sickle pod; *A. cassiae* to control Jimson weed (*Datura stramonium*) ; *C. truncatum* to control Hemp Sesbania and *F.*

laterifolium against velvet leaf (*A. theophrasti*) (Connick *et al.*, 1991) etc.

Addition of exogenous nutrients to liquid (Daigle *et al.*, 1991 and Wymore *et al.*, 1986) or granular (Boyette *et al.*, 1984 and Weidemann 1988) preparations of potential bioherbicides has shown promise in enhancing effectiveness of pathogens or increasing sporulation on the surface of granules. Enhanced efficacy has also been reported by adding surfactants and adjuvants (Grant *et al.*, 1990, Prasad 1992 Daigle *et al.*, 1990, 1991, 1992, Bannon *et at.*, 1990, Boyette *et al.*, 1991).

5   PROMINENT MYCOHERBICIDES

DEVINE : A formulation of soil borne fungus *Phytophthora palmivora* was registered in 1981 as first selective mycoherbicides for the control of a stangle vine or milk weed (*Morrenia odorata*) in Florida citrus groves, under the trade name Devine (Ridings 1986). Thus facultative parasite was reported to produce lethal root rot of host plant. Liquid suspension formulated with shelf life only six weeks was marketed as `fresh milk' and `made to order' concept, This soil applied mycoherbicide proved to be extremely effective. Kenney (1986) concluded that "the only problem that really exists with Devine is its efficacy. It is not that Devine is not active enough, but too active. Groves that were treated in 1978 to 1980 are still showing 95 to 100% control from a single treatment" .

COLLEGO: The second commercial mycoherbicide "Collego" was a formulation of an endemic anthracnose fungus *Colletotrichum glocosporioides* sp. *aeschynomene* (C.g.a.) developed to control northern jointvetch (*Aeschynomene virginica*) in rice and soybean fields of Arkansas, Louisiana and Mississippi. Dry powder formulation containing 15% spores (conidia) of C.g.a as active ingredient and 85% inert ingredients was registered in 1982 under the trade name of Collego having shelf life of 18 months. Lethal stem and foliage blight symptoms were produced on the inoculated northern jointvetch. It was first commercially available mycoherbicide for use on an annual weed in annual crops with more than 90% control efficiency (Smith 1986 and Bowere 1986).

OTHER PROMINENT MYCOHERBICIDES: These are presented in Table 1. List of other reported potential mycoherbicides which are now at the lane of research has been produced in Table 2.

6 PROSPECT:

REDUCTION OF GROWTH AND COMPETITIVENESS - AN ECOLOGICAL APPROACH

There is an increasing trend among scientists to promote the development of pathogenic micro organisms that debilitate rather than kill weeds as biological control agent. (Boyetehko *et al.,* 1993, Adcock et *al.,* 1991, Hasan *et al.,* 1990,. Kremer 1993).

INCLUSION OF MYCOHERBICIDES IN IPM PROGRAMME

It is being studied to integrate mycoherbicide into existing Pest management systems (Smith 1982b).

Interaction with other bio - agents

Synergy of fungus: There are many reports of synergy (Hasan *et al.,* 1990). Tank mixed application of C.g.a and C.g.f. sp. *Jussiacae* was observed to control *A. virginica* (96%) and *Ludwigia decurence* (100%) in rice field of Arkansas (Boyette *et al.,* 1979). Morin *et al.,* (1993) reported synergy of *Puccinia xanthii* and *Colletotrichum orbiculare* in controlling weed *Xanthium occidentale.*

Use of arthropods and fungi: Charudattan (1986) reported scope of integrated control of water hyacinth using more than one of the following biocontrol agents: fungi *Cercospora rodmani,.* weevil *Neochetina eichhorniae* and *N. bruchi,* mite *Orthogalumna terebranitis,* and moth *Arzoma dinsa* and *Someodes abligurtalis.*

Muller Scharer *et al.,* (1996) proposed system management approach for biological weed control in crops with the view of modern agroecology. It was based on the management of used pathosystem in order to maximize the material spread and disease severity of a native or naturalized pathogen. With effective biocontrol of *Senecio vulgaris* L. using the naturalized rust fungus *Puccinia laginophorae* Cooke, he concluded that biological control agents had to be seen as stress factors, not as weed killer.

Interaction of fungi with bacteria: Schister *et al.,* (1991) and Fernand *et al.,* (1994) have reported synergistic effect of epiphytic phyllosphere bacteria on *C. truncatum* in

Sesbania exaltata control and on *C. coccods* in *Abutilon theophrasti* control, respectively.

Interaction with other Chemicals

Interaction with herbicides: Numerous research work have been done to search compatibility with other herbicides as well as to increase efficacy of mycoherbicide by prior application of low dose herbicide. Smith (1986) has reported that acifluorfen or bentazon could be tank mixed with Collego but not with propanil or 2,4,5- T. Charudattan (1986) reported that pretreatment of very low dose application of 2,4,-D (6.4% of recommended dose) or paraquat (0.3% of recommended dose) had increased the efficacy of Velgo. Detailed study had been done considering C.g.m., C.g.a, *Fusarium solani* sp. *cucurbitaceae* and other mycoherbicides (Grant *et al.*, 1990, Weidemann *et al.*, 1988).

Interaction with insecticides and fungicide : Grant *et al.* (1990) while conducting detailed study of effect of pesticides on survival of C.g.m, found that tridemefon at recommended rate did not affect spore germination, but benomyl, carbathin, chlorathalonil, mancozeb, ferban, Thiram and Captan inhibited spore germination. Smith (1986) has reported compatibility of tank mixing of malathion and carbafuron with C.g.a.

Interaction with Hormone: Wymore *et al.*, (1987, 1989) have reported that a cytokinin active derivative tridiazuron had enhanced the effect of *Colletotidohm coccodes* on velvet leaf.

7   ECONOMICS

Mycoherbicides have been regarded to be cost effective. Heiny *et al.* (1993) and Zormner *et al.* (1993) have estimated that development cost regarding mycoherbicide was about 1.5 to 2 million, which is much lower than that required for development of effective herbicide formulation (about US $ 10-20 million). Bower (1982) estimated that the cost of application of Collego was about $ 29/ha.

8 SUMMERY

Bioherbicides' approach is gaining momentum. New bioherbicides will find a place in irrigated tropical, sub-tropical agro-ecosystems, forestry, waste lands as well as in managing parasite weeds or resistant weed control.

Research on synergy test of pathogens and pesticides for inclusion in IPM, developmental technology, fungal toxins, application of biotechnology, especially genetic engineering, is required. However, bioherbicides should not be viewed as a total replacement of chemicals but rather than complementary practice in integrated weed management systems.

REFERENCES:

Adcock, T.E. and Banks, P.A. 1991. Effects of pre-emergence herbicides on the competitiveness of selected weeds. *Weed Sci.* : 39: 54-56.

Andrews J.H. 1992. Biological control in the plynosphere. *Annu. Rev. Phytopathol.* 30:603-635.

Auld, B.A. 1993. Vegetable oils suspension emulsions reduce dew dependence of a mycoherbicide. *Crop Prot.* 12:477-479.

Auld, B.A., Rerae, C.F. and Say, M.M. 1988. Possible control of *Xanthium spinosum* by a fungus. *Agric. Ecosys. Environ.* 21:219-223.

Auld, B.A., Say, M.M. and Millar, G.D. 1990. Influence of potential stress factors on anthracnose development on *Xanthium spinosum J. Appl. Ecol.* 27:513-519.

Auld, B.A., Say, M.M., Ridings, H.I. and Andrews, J. 1990. Field application of *Colletotrichum biculare* to control *Xanthium spinosum. Agric. Ecosys. Environ.* 32:315-323.

Auld, B.A., Talbot, H.E. and Radburn, K.B. 1992. Host range of three isolates of *Alternaria zinniae* a potential biocontrol agent for *Xanthium. Plant Prot.* 7:114-116.

Bals, E.J. 1984. Where have all the droplets gone ? P-81-85 In Modin, R.W. (ed.) Proc. Seventh Aust. Weeds Cong., Perth.

Bannon, J.S., White, J.C., Long, D., Riley, J.A., Baragona, J., Atkins, M. and Crowby, R.H. 1990. Bioherbicide technology : an industrial perspective P. 35-319. In R.E. Hoaglard (ed.) Microbes and microbial products for herbicides. ACS. Symp. Ser. 439. American Chemical Society, Washington D.C.

Beier, R.E. 1990. Mutual pesticides and bioactive components in foods. P. 47 - 125. In: G.W. Wase (ed.) Reviews of environmental contamination and toxicology. Vol. 113 Springer Verlay. Inc., New York.

Blakeman, J.P. 1992. Fungal interaction on plant surfaces. P. 853-867. In: C.G. Carrol and D.T. Wieklow, (eds.) The fungal community its organization and role in the ecosystem, 2nd edition. Mared Dekker Inc., New York.

Bowers, R.C. 1982. Commercialization of microbial biological control agents. p. 157-173. In. R. Charudattan, and H.L. Walker, (eds). Biological control of weeds with plant pathogens. John Willey and Sons. Inc, New York.

Bowers, R.C. 1986. Commercialization of Collego - An industrialists view. *Weed Sci.* : 34 (Suppl. -1) : 24-25.

Boyetehko, S.M. and Mortensen, K. 1993. Rhifobacteria as biocontrol agents of down of brome. P - 67 (Abstract). *In* : Proc. Sixth Int. Confr. Plant Pathol., Montreal.

Boyette, C.D. 1991. Host range and virulence of *Colletotrichum truncatum* a potential mycoherbicide for hump sesbania (*Sesbania exaltata*). *Plant Dis.* 75:62-64.

Boyette, C.D. and Walker, H.L. 1989. Factors influencing biocontrol of velvetual (*Abutilon theophrasti*) and primart side (*Sida spinosa*) with *Fusarium lateritium. Weed Sci.* 33:209.

Boyette, C.D., Quimby, P.E. Jr., Bryoson, C.T., Egley, G.H. and Fulgham, F.E. 1993 a. Biological control of Hemp sesbania (*Sesbania exaltata*) under formulated in an invert emulsion. *Weed Sci.* 41:497-500.

Boyette, C.D., Quimby, P.E. Jr., Daigli, D.J. and Fulgnam, F.E. 1991. Progress in the production, formulation, and application of mycoherbicides. In : D.O. TeBeest ed. Microbial Control of Weeds. Chapman and Hall Inc. New York. p. 209-224.

Boyette, C.D., Templeton, G.E. and Oliver, L.R. 1984. Texas gourd (*Cucurbita texana*) control with *Fusarium solani* T. sp. *cucurbitae*. *Weed Sci.* 32:649-675.

Boyette, C.D., Templeton, G.E. and Smith, R.S. Jr. 1979. Control of winged wath primrose (*Jussiaca dicurrens*) and northern jointvetch *(Aeschynomene virginica)* with fungal pathogens. *Weed sci.* 27:497-501.

Boyettee, C.D., Abbos, H.K. and Connick, W.J. Jr. 1993. Evaluation of *Fusarium oxysporum* as a potential bioherbicide

46

for sickle pod (*Cassia obtusifolia,*
Sesbania exaltata); *Weed Sci.* 41:678-
681.

Bruce, A.A. and Louise, M. 1995. Constraints
in the development of Bio-herbicides,
Weed Technology 93, 638-652.

Burdan, J.J. 1987. Disease and plant
population biology; Cambridge
University Prss., Cambridge 208P.

Charudattan, R. 1986. Integrated control of
water hyacinth (*Eichhornia crassipes*)
with a pathogen, insects, and
herbicides. *Weed Sci.,* 34 (Suppl.-1)
26-30.

Charudattan, R. 1989. Assessment of efficacy
of mycoherbicide candidate P. 455-
464. In: E.S Delpossea (ed.) Proc.
Seventh Int. Symp. on Biological

Control of Weeds, Institution Sperimentale per la pathologia vegetable. Ministers dell Agricultural educe Foreste. Rome.

Charudattan, R. 1991. The mycoherbicide approach with plant pathogens P. 24-57. In D.O. TeBeest (ed.) Microbial control of weeds. Chapman and Hall Inc., New York.

Charudattan, R. and Browning, H.W. (eds.) 1992. Regulations and guidelines : Critical issues in biological control. Proc. USDA/CSRS National Workshop. Viera VA Institute of Food and Agricultural Sciences. University of Florida, Gainesville, FL.

Charudattan, R. and Walkar, H.Z. 1982. In: Biological control of weeds with plant pathogens. John Wiley and Sons, New York 293P.

Charudattan, R., Devaleria, J.T. and Prange, V.J. 1990. Special problems associated with aquatic weed control. P . 287-303. In: R.R. Baker and P.E. Dunn (eds.) New Directions in Biological Control. Alternatives for suppressing agricultural pests and diseases. Alan. R. Liss Inc., New York.

Chirsty, A.L., Herbst, K.A., Kostka, S.J., Mullen, J.P. and Carlson, P.S. 1993. Synergising weed biocontrol agents with chemical herbicides P. 87-100. In S.O. Duke, J.J. Menn. and J.R. Plimmer (eds.) Pest control with enhanced environmental safety. ACS

Symp. Ser. 524 American Chemical
Society, Washington DC.

Combellack, J.H. 1981. An assessment of the
problems of effectively spraying
herbicides onto weeds in cropped
areas. P. 93-97 (Vol.1) In: J.T. S.
Warbrick and B.J. Wilson (eds.) Proc.
Sixth Aust. Weeds Cong., Gold Coast.
Queensland.

Connick, Jr. W.J., Boyette, C.D. and Mc
Alpine, J.R. 1991. Formulation of
mycoherbicides using a posta-live
process. *Biol. Control* 1:281-287.

Connick, Jr. W.J., Lesi, J.A. and Quimby, P.C.
Jr. 1990. Formulation of biocontrol
agents for use in plant pathology. P.
345-372. In R.R. Baker and P.E. Dunn.
(eds.) New directions in biological

control. Alternatives for suppressing agricultural pests and diseases. Alan. R. Liss Inc., New York.

Cunnigham, J.E. and Kuiack, J.E. 1989. Esterase activity as a market for sporulation in *Colletotrichum glocosporioides* T. sp. *malvae* in submerged culture. *Mycol. Res.* 93:236-239.

Daigle, D.J. and Cotty, P.J. 1991. Factors that influence germination and mycoherbicidal activity of *Alternaria cassiae. Weed Technol.* 5:82-86.

Daigle, D.J. and Cotty, P.J. 1992. Production of conidia of *Alternaria cassiae* with alginate pallets *Biol. Control* 2:278-281.

Daigle, D.J., Connick, W.J. Jr., Quimby, P.E. Jr., Evans, J., Trosk-Morrell, B. and Fulgnam, F.E. 1990. Invert emulsion : carried and water source for the mycoherbicide *Alternaria cassiae. Weed Technol.* 4: 327-331.

Deverall, B.J. 1977. Defense mechanisms of plants. Cambridge University Press. Cambridge 100P. Feicthenberger, E. Zentmyer, G.A. and Menge, J.A. 1984. Identity of *Phytophthora* isolated from milk weed vine. *Phytopathology* 74:50-55.

Grant, N.T., Orusin, P. Kiewiet, E., Makowski, R.M., Holmstrom, D., Ruddick, B. and Mortensen, K. 1990. Effect of selected pesticides on survival of *Colletotrichum glocosporioides* L. sp. *malvae,* a bioherbicides for round named Mallow (*Malva pusilla*) 4:761-715.

Greaves, M.P. and Maequeen, M.D. 1992. Bioherbicides : their role in tomorrow's agriculture. P. 295-306. In: L. Denholm, A.L. Devonshire (eds.) Developments in combating pesticide resistance. Elsevier Applied Science. London.

Gressel, J., Amsellem, Z., Warshawasky, A., Kampel, V., Miechael D. 1996. Biocontrol of weeds : overcoming evolution for efficacy. *Journal of Environmental Science* ; B.31: 399-405.

Grover, R. 1991. Nature of transport and fate of air borne residues. p. 89-117. In: R. Grover and A.J. Cessna (eds.)

Environmental chemistry of herbicides;
Vol. 2 CRC Press. Inc. Boea Raton,
FL.

Hasan, S.P.G. and Ayres, P.G. 1990. Tansky
review no. 23. The control of weeds
through fungi. Principles and
Prospects. *New Phytol.* 115 : 201-222.

Heinly, D.K. and Templeton, G.E. 1993.
Economic comparisons of
mycoherbicides for conventional
herbicides. p. 395-408. In : J. Altman
(ed.) Pesticide interaction in crop
production : Beneficial and deleterious
effects. CRC Press Inc. Boca Raton.
FL.

Hesseltine, C.W. 1986. Global significance of
mycotoxins. p. 1-18. In: P.S. Steyn and
R. Vieggear (eds.) Mycotoxins and

phytotoxins. Elsevier Science Publishers. Amsterdam.

Holt, J.S., Powles, S.B. and Hotherm, J.A.M. 1990. Carbon concentration and carbon to nitrogen ratio influence submerged culture condition by the potential bioherbicide *Colletotrichum truncatum*. NRRL 13737; *Appl. Environ. Microbial* 56:3435-3438.

Jackson, M.A. and Bothast, R.J. 1990. Carbon concentration and carbon to nitrogen ratio influence submerged culture condition by the potential bioherbicide *Colletotrichum truncatum*. MRRL 13737. Appl. *Environ. Microbiol.* 56: 3435-3438.

Jackson, M.A. and Schtister, D.A. 1992. The composition and attributes of

Colletotrichum truncatum spores are altered by the natural environment. *Appl. Environ. Microbiol.* 58:2260-65.

Jones, R.W. and Hanockc, J.G. 1990. Soil borne fungi for biological control of weeds. P - 276-286. In: R.E. Hoagland (ed.) Microbes and microbial products as herbicides. ACS Symp. Ser. 439. American. Chemical Society, Washington Dc.

Kenerley, C.M. and Andrews, J.H. 1990. Interactions of pathogens on plant surfaces. p. 192-217. In: R.E. Hoagland (ed.) Microbes and microbial products or herbicides. ACS Symp. Ser. 439. American Chemical Society. Washington. Dc.

Kenney, D.S. 1986. Devine the way it was developed-an industrialists view, *Weed Sci.* 34 (suppl. 1) 15-16.

Kremer, R.J. 1993. Management of weed seed banks with microorganisms. *Ecol. Applic.* 3 : 42-52.

Lym, R.G. Messersmith, C..G 1988. Survey picloram in Horth Dakota groundwater *Weed Techn.* 38:217-222.

Makowski, R.M.D. 1993 a. Effect of inoculum concentration, temperature, dew period, and plant growth stage on disease of round leaved mallow and velvet leaf by *Colletotrichum glocosporioides* sp. *malvae* *Phytopathology.* 83:11, 1229-1234.

Makowski, R.M.D. 1993. Foliar pathogens in weed biocontrol : Ecological and

regulatory constraints. P. XX -XX. In:
D.A. Andow, D.W. Ragsdale and R.F.
Nyvall (eds.) Wesview Press. Boulder
Co.

Makowski, R.M.D. and K. Mortenson, 1992.
The first mycoherbicide in Canada :
*Colletotrichum glocosporioides T. sp.
malvae* for round leaved mallow
control. P. 298-300 in R.G. Richardson.
(ed.) Victoria Inc., Melbourne. Victoria.

McRae, C.F., Riddings, H.I. and Auld, B.A.
1988. Anthracnose of *Xanthium
spinosum-* qualitative disease
assessment and analysis. *Aust. J.
Plant Pathol.* 17 : 11-13

Merae, C.F. and Auld, B.A. 1988. The
influence of environmental factors on

anthracnose of *Xanthium spinosum.*
Phytopathology 78:1182-1186.

Mikandrow, A., Weidemann, G.J. and Auld,
B.A. 1990. Incidence and pathogenicity
of *Colletotrichum orbiculare* and
Phomopsis on *Xanthium* sp. *Plant Dis.*
74:796-799.

Misaghi, I.J.1982. Physiology and
biochemistry of plant pathogen
interactions. Plenum Press. New York.
P. 287.

Morin, A., Watson, A.K. and Reelder, R.D.
1990a. Effect of dew, inoculum density
and spray additives on infection of field
bindweed by *Phomopsis convoluvlus.*
Can J. Plant Pathol. 12:48-56.

Morin, L. Watson, A.K. and Reelder, R.D.
1990 b. Production of conidia of

Phomopsis convoluvlus. Can. J. Microbiol 36:86-91.

Morin, L., Auld, B.A. and Brown, J.F. 1993. Synergy between *Precimia xanthii* and *Colletotrichum orbiculare* on *Xanthium occidentale; Biol. Control* 3 : 296-310.

Morin, L., Watson, A.K. and Reeleder, R.D. 1989. Efficacy of *Phomopsis convoluvlus* for control of field bind weed (*Convoluvlus arvensis*). *Weed Sci.* 37 : 830-835.

Moss, B.R. and Rubin, B. 1993. Herbicide resistant weeds : a worldwide perspective. *J. Agric. Sci.* 120 : 141-148.

Muller - Scharer, H. and Frantfen, J. 1996. In emerging system management approach for biological weed control in crops : *Senecio vulgaris* as a research model. *Weed Research* 36: 483-491.

Pimental, D. and Levitan, L. 1986. Amount applied and amount reaching pests. *Bioscience* 36: 86-91.

Poul, N.D., Ayres, P.G. and Hallett, S.G. 1992. Making biological herbicides more effective; *J. Biol Educ.* 26:94-99.

Prasad, R. 1992. Role of adjuvants in modifying the efficacy of a bioherbicide on forest species. Compatibility studies under laboratory conditions. In: 3rd Int. Symp. Adjuvants for Agrochemicals. 3-7 Aug. 1992. *Pesticide Sci.* 38 : 2-3, 278-275.

Quimby, Jr. P.E., Fulgham, F.E., Boyette, C.D. and Connick, W.J. Jr. 1989. An invert emulsion replaces dew in biocontrol of sicklepod a preliminary study. P. 264-270. In: D.A. Hovde and G.B. Beestman, (eds.) Pesticide formulations and application system. Vol. 8 ASTM STP 980. American Society for Testing and Materials, Philadelphia.

Raj Prasad. 1993. Development of bioherbicides for integrated weed management in forestry. In: Proc. Int. Symp. Indian Society of Weed Science. Hisar, Nov. 18-20, Vol. 1 : 269-274.

Rounkiar, C. 1983. Five forms of plants and statistical plant geography. Calendor Press. Oxford. 632 P.

Saliwanchik, R. 1986. Patenting/licensing of microbial herbicides. *Weed Sci.* 34 (Suppl. 1) 43-49.

Saliwanchik, R. 1988. Protecting biotechnology inventions : a guide for scientists. *Science Tech.* Publishers. Madison, WI. 175P.

Schepens, P.C. 1987. Joint action of cachio botus puratus and abasine on *Echinochloa crus-galli* (L). Beav. *Weed Res.* 27:43-47.

Schister, D.A., Jackson, M.A. and Bothast, R.J. 1991. Influence of nutrition during conidiation of *Colletotrichum truncatum* on conidial germination and efficacy in inciting disease in *Sesbania excellata Phytopathology* 81:587-590.

Schoenewiess, D.F. 1975. Predisposition. Stress and plant disease. *Ann. Rev. Phytopathol.,* 13:193-211.

Smith, A.E. 1982a. Herbicides and the soil environment in Canada. *Can. J. Soil. Sci.* 62:433-460.

Smith, R.J. Jr. 1982b. Integration of microbial herbicides with existing pest management programs. p. 189-20. R. Charudattan and H.L. Walker (eds.) Biological control of weeds with plant pathogens. John Willey and Sons, New York.

Smith, R.J. Jr. 1986. Biological control of northern joint vetch (*Aeschynomene virginica*) in rice (*Oryza sativa*) and soybean (*Glycine max*) - A researcher's view. *Weed Sci.* 34: (Suppl. 1) 17:23.

Strobel, L.J. 1991. Submerged fermentation of biological herbicides P. 225-261. In: D.O. TeBeest (ed.) Microbial control of weeds. Chapman and Hall Inc., New York.

TeBeest, D. and Templeton, G. 1985; Plant disease 61:6 -10.

TeBeest, D.O. 1991. Ecology and epidemiology of fungal plant pathogens studied as biological control agents of weeds. P. 97-114 In: D.O. TeBeest (ed.) Microbial control of weeds. Chapman and Hall Inc., New York.

TeBeest, D.O., Cisar, C.R. and Spiegel, F.W. 1992 a. Partial characterization of progeny from a cross between

Colletotrichum glocosporioides sp. aeschynomene and C. *glocosporioides* from *Cana. Plant Prot.* 7:171.

TeBeest, D.O., Yang, X.B., and Cisar, C.R. 1992b. The status of biological control of weeds with fungal pathogens. *Annu. Rev. Phytopathol.* 30:637-657.

Templeton, G.E. 1982 b. In: Biological control of weeds with plant pathogens. R. Charudattan and R. Walker (eds.) John Willey and Sons, N.Y. pp 29-44.

Templeton, G.E. 1982. Biological herbicides: Discovery, development, deployment. *Weed Sci.* 30:430-433.

Templeton, G.E. 1992. Some `orphanod' mycoherbicides and their potential for development. *Plant Prot.* 7:149-150.

Templeton, G.E. 1992. Use of *Colletotrichum* strains as mycoherbicide. P. 358-380. In: J.A. Balley and M.J. Jager (eds.) *Colletotrichum* : Biology, Pathology and Control ; British Society for Plant Pathology. C.A.B. International Walling Ford.

Tukey, H.B. 1970. The leaching of substances from plants. *Annu. Rev. Plant. Physiol.* 21:305-324.

Vanderplank, J.E. 1975. Principals of plant infection. Academic Press, New York. 216 P.

Walker, H.L. 1981. Granular formulation of *Alternaria macrospira* for control of spurred anoda *(Anoda cristata). Weed Sci.* 29:342-345.

Walker, H.L. and Conick, W.J. Jr. 1983. Sodium alginate for production and formulation of mycoherbicides. *Weed Sci.* 31:333-338.

Walter, J., Mikandrow, A. and Mittar, G.D. 1991. Species of *Colletotrichum* on *Xanthium* (Asteraceae) with comments on some taxonomic and nomenclatural problems in *Colletotrichum. Mycol. Res.* 95:1175-1193.

Watson, A.K. 1989. In Proc. Bright Crop Protection Cant. *Weeds* 987-996.

Weidemann, G.J. 1988. Effects of nutrition amendments on conidial production of *Fusarium solani* T. sp. *cucurbitae* on sodium alginate granules and on control of Texas gourd. *Plant Dis.* 72:757-759.

Weidemann, G.J. 1991. Host range testing : safety and science. P. 83-96 in D.O. TeBeest (ed.) Microbial control of weeds. Chapman and Hall Inc., New York.

Weidemann, G.J. 1992. Risk assessment determining genetic retardness and potential asexual gene exchange in biocontrol fungi. *Plant Prot.* 7:166-168.

Weidemann, G.J. and TeBeest, D.O. 1990. Genetic variability of fungal pathogens and their weed hosts. In: R.E. Heogland (ed.) Microbes and microbial products as herbicides. ACS Symp. Ser. 439. American Chemical Society. Washington. Dc. p - 176-183.

Weidemann, G.J. and Templeton, G.E. 1988 . Control of Texas gourd *Cucurbita taxana,* with *Fusarium solani* T. sp. *cucurbitaceae. Weed Tech.* 2:271-274.

Windels, C.E. and Lindow, S.E. 1985. Biological control on the phylloplane. American Phytopathological Society. St. Paul. MM.

Womack, J.G. and Burge, M.N. 1993. Mycoherbicide formulation and the potential for bracken control. *Pesic Sci.* 37:357-341.

Wymore, L.A. and Watson. A.K. 1986. An adjuvant increases survival and efficacy of *Colletotrichum coccodes.* A mycoherbicide for velvet leaf (*Abutilon theophrasti*) *Phytopathology* 76:115-116.

Wymore, L.A., Poiries, C., Watson, A.K. and Gotlieb, A.R. 1988. *Colletotrichum coccodes* a potential bioherbicide for control of velvet leaf (*Abutilon theophrasti*). *Plant Dis.* 72 : 534 538.

Wymore, L.A., Watson, A.K. and Gotlieb, A.R. 1987. Interaction between *Colletotrichum coccodes* and Tridiazuron for control of velvet leaf (*Abutilon theophrasti*). *Weed Sci.* 35:377-383.

Wymore, L.A., Watson, A.K. and Gotlieb, A.R. 1989. Interaction between *Colletotrichum coccodes* and *Tridiazuron* for velvet leaf (*Abutilon theophrasti*) control in the field, *Weed Sci.* 37: 478-483.

Yang, X.B.A and TeBeest, D.O. 1993. Epidemiological mechanisms of mycoherbicide effectiveness, *Phytopathology* 83:891-893.

Zorner, P.S., Evans, S.L. and Savage, S.D. 1993. Perspectives on providing a realistic technical foundation for the commercialization of bioherbicides. P. 79-86. In: S.O. Dune, J.J. Menn, and J.R. Plimmer (eds.) Pest control with enhanced environmental safely. ACS Symp. Ser. 524. American Chemical Society, Washington DC.

Table 1 Prominent mycoherbicides

SI. No.	Fungal agent	Trade name	Target weed(s)	Remarks
1	*Alternaria cassiae*	Casst	*Cassia obtusifolia* *C. occidentalis* *C. alata*	Used in soybean, cotton and peanut in Florida
2	*A. macrospora*		*Anoda cristata*	Applied in cotton
3	*Cercospora rodmani*	ABG-5003	*Eichhornia crassipes*	In water ways of USA
4	*Colletotrichum glocosporioides sp. Cuscutae*	LUBAO	*Cuscuta* sp.	China
5	*C. g.f.sp. malvae*	Biomal No 1 & 2	*Malva pusilla*	On wheat and small grain crops in USA
6	*C. coccodes*	Velgo	*Abutilon theophrasti*	—

Table 2. A sampling of pathogens used, in the market or amid study

SI. No.	Pathogen	Weed	Where Weed is troublesome	Source
Now sold commercially				
1	*Colletotrichum glocosporioides*	Northern joint vetch *Aeschynomene virginica*	Rice crop in Arkansas, USA	Strobel, 1991 Daniel *et al.* 1973
2	*Phytophthora demivora* Trade name: Devine	Milk weed vine *Morrenia odorata*	Citrus orchards in south east USA	Strobel, 1991 Feichtenberger *et al.*, 1984
Expected to be marketed				
3	*Alternaria cassiae*	Sickle pod *Cassia obtusifolia*	Soybean and peanut crop in Southern USA. Various crops in Australia.	Ayres and Peu, 1990
4	*Colletotrichum glocosporioides* Trade Name: Biomal	Round leaf hollow *Malva pusilla*	Wheat and small grain crops in USA	Strobel, 1991 and Auld, *et al.* 1990

Showing promise				
5	Colletotrichum orbiculare	Xanthium spinosum	Australia	Auld et al., 19
6	Colletotrichum glocosporioides	Heke serices	Sucope province South Africa	Mursis, 1989
7	Colletotrichum truncatum	Hecurp sestbies Sesbania exallets	Mississippi USA in soybean crop	Boyette, 1991
8	Colletotrichum glocosporioides	St. John's wort Hypericum perforatum	Australia	Hildenbrend ar Jessen 1991 Shepane 1995
9	Puccinia jaceae Certaures solstialis P. aonoptilie P. aermiahensis	Yellow star	North America At pasture land of USA. Cultivated land of Canada	Watson et al., 1986
10	Botrutis cinera	Senecio vulgarica	Groundnut	Hellett et al.,
11	Puccinia logenophorae	-do-	-do-	-do-
12	Rumularia subells	Rumex obtusifolius		Huber Meizicke et al., 1989
13	Exerohitum Turcicum	Johnson Grass (Sorghum halepense)		Chiang et al., 1989

14	*Colletotrichum graminicola*	-do-	-do-	-do-
15	*Glococerspora sorghi*	-do-	-do-	-do-
16	*Bipodaris relepense*	-do-		
17	*Nectria galligena*	*Myrica feys*	Azores coney Island at Hawaii	Gordner and Hadges, 1990
18	*Bortyosphe- riaribis* spp.	-do-		
19	*Cryphonectoa* spp.	-do-		
20	*Ranuleris destructive*	-do-		
21	*Alternaria cassiae Alteruris cressa*	Cassia	Israel	Ami Sellem *et al.*, 1990
22	*Alteruris cressa*	*Datura stramonium*	Israel, USA	Boyette *et al.*, 1991
23	*Metschnikowia seukanji*	Field milk weed *Asclepias syriaca*	USA	Eisikowitch *et al.*, 1990
24	*Septoria* spp.	*Convoluvlus arvensis*	Greece in Maize and potato	Giannopolits *et al.*, 1989
25	*Phomopsis convoluvlus*	-do-	USA	Morin *et al.*, 1989,1990

26	*Fusarium monifliorme*	Jimson weed (*Datura stramonium*)		Abbus *et al..*, 1991
27	*Alternaria deternota*	*Arctostaphylos Pethula*	North India	Aneja & Singh 1989 Elwakil *et al..*, 1990
28	*Dredstera* spp.	*Costenopsis velutina*	Oregon, USA	Crowford *et al* 1987
29	*Truncetellaangusteta*	*Cegnothus velutina*		Joye, 1990
30	*Macrophomina phaseolina*	*Hydrilla verticillata*	Texas, USA	Schepens, 198
31	*Chondrostercum purpureum*	*Prunus serotina*	*Netherlands*	
32	*Melanconca* spp.		*Alnus rulra*	
33	*Hypoxylon mansumatuam*		Canada	
34	*Nectria ditissiam*			Dorworth, 199
35	*Phyllosicta pyrolae*	*Gaultheria shallon*		
36	*Glindrocarpon*	*Rubus cunefolius*		
37	*Chondrosterum purpureum*	Broadleaved trees		
39	*Cochliobolus corbonum*	*Euphoriba geniculata*	South Zamiaba	Lakshumna *et al.*, 1990

39	*Gliodadium virens*	Broad spectrum control		Jones and Hanockc, 1990
40	*Uromyclodium teppericnyum*	*Acodia saligan*	South Africa	Plant Protection, 1987
41	*Colletotrichum coccades*	*Abutilon theophrasti* eastern black night shade *Solanum candinum*	Canada	Godoury and Watson, 1987

www.ingramcontent.com/pod-product-compliance
Lightning Source LLC
Chambersburg PA
CBHW021015180526
45163CB00005B/1959